# Edible Mushrooms are marked with a GREEN circle and a WHITE number.

 **Edible mushrooms** – *mushrooms for beginners* – that can only be confused with other edible mushrooms.

 **Edible mushrooms** – *mushrooms for beginners* – that can be confused with mushrooms unsuitable for consumption for reasons other than being poisonous.

 **Edible mushrooms** that can be confused with mildly poisonous mushrooms.

 **Edible mushrooms** that can be confused with dangerously poisonous mushrooms.

Edible mushrooms with certain restrictions are marked with a **black ring** around the green circle in the following way:

 **Edible mushrooms** that can only be confused with other edible mushrooms and that must first be blanched in plenty of water (see p. 94–95).

 **Edible mushrooms** that can be confused with mildly poisonous mushrooms and that must first be blanched in plenty of water (see p. 94–95).

# Inedible mushrooms are marked with a **RED** or YELLOW square.

## Poisonous Mushrooms (marked in red)

A large proportion of mushrooms contain self-produced toxins that are poisonous. A mushroom containing dangerous levels of these toxins is known as a poisonous mushroom.

☠ *Dangerously poisonous mushrooms,* especially those containing substances damaging to cells or large amounts of substances damaging to the nervous system, will be marked with *a red poison symbol.*

## Mushrooms unsuitable for consumption due to reasons other than being poisonous

Most of our larger mushrooms belong to this group. The look-alike mushrooms that are marked in this way in yellow belong to one or more of the following categories:

1. **Mushrooms lacking sufficient research.**
2. **Inedible mushrooms.**
3. **Worthless mushrooms.**
4. **Edible but rare mushrooms.** (Please note that these are regarded as edible mushrooms in countries where they grow abundantly.)

# The Pocket Guide to Wild Mushrooms
## Helpful Tips for Mushrooming in the Field

Pelle Holmberg    Hans Marklund

Translated by Ellen Hedström

Skyhorse Publishing

Skyhorse Publishing books may be purchased in bulk at special discounts for sales promotion, corporate gifts,
fund-raising, or educational purposes. Special editions can also be created to specifications. For details, contact the Special
Sales Department, Skyhorse Publishing, 307 West 36th Street, 11th Floor, New York, NY 10018 or info@skyhorsepublishing.
com.

Skyhorse® and Skyhorse Publishing® are registered trademarks of Skyhorse Publishing, Inc.®, a Delaware
corporation.

Visit our website at www.skyhorsepublishing.com.

13

Library of Congress Cataloging-in-Publication Data is available on file.
ISBN: 978-1-62087-731-9

Cover design: Seth Kapadia; cover photograph Hans Marklund (front cover; above
**chanterelle,** below **orange oak bolete, black morel, and yellow foot**) and Ingrid Holmberg
(author's portrait on the back cover)

Printed in China

# Contents

*Herald of the winter*
one of late fall's delicacies

# Introduction

The Pocket Guide to Wild Mushrooms is a smaller, portable version of our book Nya Svampboken, to be used in the field. The selection of species included is reduced, however it still includes the most important edible mushrooms and of course the most common look-alike mushrooms. Those who would like further information about the mushrooms presented in this book or find some species missing can refer to other resources, listed in the back of the book.

In the first edition of Nya Svampboken (1996), we introduced new symbols for the mushrooms and these can also be found in this book. The symbols are based on each edible mushroom's potential look-alike counterpart. Edible mushrooms in group 1 can only (within reason) be confused with other edible mushrooms. Those in group 2 can be confused with mushrooms that are unsuitable for consumption but that are not poisonous. Group 1 and 2 are what we call beginners mushrooms. Edible mushrooms in group 3 can be confused with slightly poisonous mushrooms, and those in group 4 can be confused with lethally poisonous mushrooms. We consider all mushrooms presented here to taste good, but which taste the best? Well it depends on a lot of factors including how they are prepared, what they accompany, including beverages, and of course personal taste.

Have fun foraging for mushrooms!

Pelle Holmberg        Hans Marklund

# What is a mushroom?

All mushrooms presented here are made of **mycelium**, a network of thin threadlike structures in the ground (see photo on right). The mycelium forms **fruiting bodies** (see below) that mature and become attractive to mushroom pickers. The fruiting bodies can be compared to the fruit on a tree and have as their biological task to form spores, or **reproductive bodies**.

*White and yellow **mycelium** underneath a decaying log on the ground.*

Cap

*Scales,* remnants of a **universal veil**

*Ring,* remnants of a **partial veil**

*Stipe* with a reticular pattern

**The fruiting body of a Parasol Mushroom**

*Remnants of a **universal veil** that has covered the entire button. It then forms a **volva** around the base and **patches or warts** on the cap.*

**Young fruiting body (button) from a mushroom of the Amanita genus**

*Remnants of a **partial veil** that forms a **ring** around the stipe.*

# When can mushrooms be found?

In order for the mycelium to develop fruiting bodies, they need rain. A dry summer usually means a lack of mushrooms; but if a dry summer is followed by a lot of rain, it can then produce a fall rich in mushrooms. A summer and fall with both sun and rain give good odds for a bountiful mushroom season. A rainy summer can produce a lot of mushrooms but if there is too much rain, the soil can become sodden, resulting in fewer mushrooms to pick. Each season is unique; a year of trumpet chanterelles precedes a penny bun year and so on. Some years August may be the best month for mushrooms and another year may find that September-October give the biggest yield. Other years the middle of the country may have a fantastic harvest whereas the north may fare less well.

The variations are significant. "Normal" seasons seem to become less frequent when it comes to the weather but usually August and September are the big mushroom months. On the west coast of the US, the season extends into the winter months, the time of best rain. States in the Southwest and Southeast US enjoy mushrooming well into the early winter months.

# Some tips regarding...

## picking

Mushrooms are fresh produce and therefore **fragile.** Some mushrooms such as boletes, brittlegills, and the shaggy ink cap have to be taken care of as soon as you get home. Others such as chanterelles and those belonging to the craterellus genus are less sensitive but should be stored in a cool place if they are not to be used right away.

Use a **basket** rather than a plastic bag when picking mushrooms. The moisture from the mushrooms can't evaporate through plastic and the warm air will create a greenhouse effect. This allows bacteria to thrive making the mushrooms quickly become inedible.

## cleaning

Clean away any soil and leaf litter from the mushrooms at the picking stage. Larger mushrooms can be halved to check if they have been contaminated by maggots. Remove any parts infected by maggots or slugs, for instance the earthy part of the stipe, and, with larger boletes, the soft pore surface.

12

Once you've returned home, place the mushrooms on a table. This is especially important if the mushrooms are damp. Feel free to **organize** them right then and there, placing spicy and strong tasting mushrooms with each other. For example the chanterelle, look-alike saffron milk cap, weeping milk cap, and the crab brittlegill should be divided up. They can then add flavor to milder tasting mushrooms.

## preparation

All mushrooms that are to be eaten must be cooked. This includes *heating* the mushrooms and allowing them to release their water. Previously some varieties have been recommended for use as raw food but later research has shown that raw mushrooms contain heat sensitive substances that can be harmful to humans. Preparation can be separated into two stages: **1. pretreatment**, either *blanching,* e.g. red hot milk cap (p. 94 and 119), or *precooking* ("to cook in its own juices") and **2. cooking**.

You can easily **preserve** mushrooms by *drying.* Tips for preparing and preserving mushrooms are given within this book for each species. (You can also refer to *Nya Svampboken* or other mushroom guides.)

13

# What do mushrooms contain?

Around 10 percent of the mushrooms fruiting body is made up of dry substances which are mainly fibrous cell walls, nutrients, vitamins, and minerals. The other 90 percent consists of water.

## Fiber

The cell walls of a mushroom are not made up of cellulose like some plants, but another form of carbohydrate that is similar to chitin, which is found in the shells of crustaceans and insects. It is likely that this chitin has a similar positive effect on our intestinal system as vegetable fiber.

## Nutrients

Mushrooms contain little or no amount of proteins, fats, or carbohydrates. The *proteins* that can be found in mushrooms are mainly enzymes that contain vital amino acids and are of a better quality than many other plant proteins. In 3.5 oz (100 g) of fresh mushrooms there is only a small amount of protein (a few grams). The amounts of fat and digestible carbohydrates are insignificant in terms of nutritional value. Small amounts of sugar, especially mannitol, are present in mushrooms and this contributes to the taste. Therefore mushrooms are very low in calories and even if you add fat during cooking, the dish will not be especially fatty. *Adding fat during preparation will however enhance the flavor as the flavors in mushrooms are fat soluble!*

## Vitamins

There are some vitamins in mushrooms but only vitamin B and D are of significant amounts. Vitamin C can be found in many mushrooms in quantities comparable to those in berries and vegetables but, as mushrooms need to be well cooked, heat sensitive vitamins such as vitamin C will mainly be destroyed.

## Minerals

It might not sound like much, but the fact that minerals make up 1 percent of the mushrooms fresh weight is actually a significant amount. Deer, moose, and other animals eat a lot of mushrooms, which is likely due to the high amount of minerals. Stories colloquially told by elders spoke of cows behaving strangely during fall when mushrooms started to appear. They became spellbound and ran into the woods, where they would eat copious amounts of mushrooms. Why then were cows in the old days so prone to eating mushrooms? Today's well fed cows appear to show less interest, probably because they get their mineral needs satisfied by their fodder. Apart from vast amounts of potassium, many mushrooms also contain good amounts of calcium, zinc, iron, manganese, and selenium.

## Conclusion:

What makes mushrooms interesting food is their role in adding flavor as well as the amount of fiber, important minerals, antioxidants, and vitamins B and D that they contain.

# An explanation of the varieties

**Birch bolete**

## Introduction

We have chosen to present fifty or so types of mushrooms that we feel are suitable for cooking in one way or another. With a few exceptions, each mushroom is presented on a page with a professional studio shot image, as well as an image taken in nature, with accompanying words.

## Description of the species

- When it comes to the actual **mushroom names,** for both the common and formal scientific versions, we have in the first instance used *Svenska svampnamn,* by Nils Lundqvist and Olle Persson. Older names, of which many are still in use, have been given in italics.

- **Reference to smell.** For a mushroom picker, a sense of smell is of equal importance to that of a wine connoisseur. Most mushrooms have a special odor, but what does that entail? It might be pleasant for one but on the contrary unpleasant for another. Our opinions on smell are based on our own experiences as well as what we have found the general consensus to be.

- **Reference to taste.** When we state the *flavor* under distinguishing features we mean the flavor of the *raw mushroom*. NOTE! With the exception of brittlegills (p. 96), the flavor of the mushroom tells us nothing of its edibility!

- Concerning the texts on species, as in the rest of the book, there are words and phrases that are not explained. Rather than having a chapter with **definitions**, we refer to the **index**, located at the end of the book, where we give the page number of the word that is being highlighted. Symbols, as well as some phrases that are specific for that variety, are explained on the following page.

## Artwork

*Saffron milk cap*

We present two types of pictures; ones taken in nature and ones taken in a studio. Past books show mushrooms in their natural habitat which is the classical way to photograph mushrooms. The problem with that, however, is it can be difficult to show the accurate coloring of the mushroom as well as to detail the fruiting bodies' various appearances within the same species. This can be avoided by photographing the mushrooms in a studio. If using artificial light, e.g. a flash, it gives the most even lighting as well as the most accurate description of the mushrooms color.

# Terms and symbols

**Edible Mushrooms** are marked with a **GREEN** circle and a WHITE number.

## Edible mushrooms

With *edible mushrooms* we imply a species that is able to be used in cooking. An important note: the mushroom should contain *flavor* that, after preparation, is perceived as pleasant or interesting.

*Saffron milk cap*

## Related edible mushrooms

suggests that the species is systematically related to the edible mushroom.

## Look-alike Mushrooms

A mushroom that can, through its appearance, *easily* be confused with an edible mushroom is called a *look-alike mushroom*. Look-alike mushrooms can either be edible mushrooms or mushrooms that should be avoided due to being poisonous or unsuitable for other reasons.

## Risks of "Look-alike Mushrooms"

When a look-alike mushroom is unsuitable for consumption (see following page) there is often a *risk* associated with picking and eating the edible mushroom. For this reason we have classified edible mushrooms into four *"risk zones"* (using the numbers 1–4):

 **Edible Mushrooms** – *beginner mushrooms* – that can only be confused with other edible mushrooms.

 **Edible Mushrooms** – *beginner mushrooms* – that can be confused with mushrooms that are unsuitable to eat for reasons other than being poisonous.

 **Edible mushrooms** that can be confused with dangerously poisonous mushrooms.

 **Edible mushrooms** that can be confused with dangerously poisonous mushrooms.

Edible mushrooms that have certain restrictions are marked with a black ring around the green circle in the following way:

 **Edible mushrooms** that can only be confused with other edible mushrooms and which must first be blanched in plenty of water (see p. 94–95).

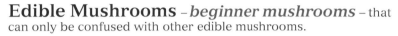 **Edible mushrooms** that can be confused with mildly poisonous mushrooms and which must first be blanched in plenty of water (see p. 94–95).

# *Inedible mushrooms* are marked with a **RED** or YELLOW square.

## Poisonous Mushrooms

A large proportion of mushrooms contain self-produced toxins that are poisonous and can cause a variety of forms of distress. A mushroom containing dangerous levels of these toxins is known as a poisonous mushroom.

☠ *Dangerously poisonous mushrooms,* especially those containing substances damaging to cells, or large amounts of substances damaging to the nervous system, will be marked with *a red poison symbol.*

## Mushrooms unsuitable for consumption due to reasons other than being poisonous

Most of our larger mushrooms belong to this group. The look-alike mushrooms that are marked in this way in yellow belong to one or more of the following categories:

1. **Mushrooms lacking sufficient research.**
2. **Inedible mushrooms.**
3. **Worthless mushrooms.**
4. **Edible but rare mushrooms.** (Please note that these can be regarded as edible mushrooms in countries where they grow abundantly).

20

*Ready set go!*

# A presentation of the species

## Edible mushrooms

*Bloody milk cap*

# Poisonous mushrooms

*Deadly web cap*

## Ignorance and Carelessness

Sometimes incorrect information states that mushroom poisoning has occurred due to mistaken identity. Often it is due to the picker **lacking knowledge** regarding poisonous mushrooms and in some cases being careless about not looking closely enough at each fruiting body, which can lead to mixing poisonous mushrooms with edible mushrooms.

## NOTE!

If you find a mushroom that does not fit any of the descriptions of the edible mushrooms described in this book, you should not try it. The risk is far too high that it could be unsuitable for consumption, as many of the most common edible mushrooms that grow in cooler climates are listed in this book. Remember, there are several thousands of different mushrooms in North America and almost as many in the UK!

# **4** Black morel *Morchélla eláta*

1cm

# Black morel

When you discuss the morel as a delicacy in the US, most refer to the Yellow Morel; but the **black** morel is another equally desirable variety of species belonging to the genus Morchella. It is vital that you distinguish these from the dangerous false morel (see below) that is eaten in a few European countries but considered dangerously toxic in the US and UK.

In the US and UK, there are several types of morels, and all are decent edible mushrooms, even if the taste can vary a great deal. The Black Morel is most common in more northern forests. It produces fruiting bodies in mid-spring but its appearance can vary from year to year.

Black morels are saprophytes, which indicates that they thrive on dead organic material such as decaying plants. They can grow in both in deciduous and coniferous woods as well as on cultivated land.

## Distinguishing features
- The cap is covered in ridges and pits.
- The outside of the cap has a very typical reticular pattern with *deep grooves* between raised *vertical ridges*. The bottom of the cap is attached to the stipe.
- The hat and stipe are hollow.

## Preparing and preserving
When it first became known that the false morel contained a deadly poison, which mostly disappears when cooked, it subsequently seemed natural to also recommend that the black morel be blanched. Today we know that the black morel *does not* need to be blanched, as neither the toxic elements in the false morel nor any other hazardous substances can be found in the black morel. Black morels work well in stews, sauces, and soups. It can be preserved in different ways but works especially well when dried.

## Look-alike mushrooms
Black morels can be confused with other types of morels, among others the **false morel**, *Gyrómitra esculénta* (photo p. 116). Regarded by many as an edible mushroom, it is the most dangerous as it contains a carcinogen damaging to the central nervous system known as gyromitrin. (For further information see p. 114–116.

1cm

# Black trumpet

A highly regarded edible mushroom which is easy to recognize but hard to find among the leaves and grass.

Commonly found in mixed hardwood and conifer forests and especially fond of moist beech forests, it can also be located in mossy coniferous woods. The black trumpet fruits in trooping clusters; if you find one, carefully seek out more in the same area. Common across North America and widespread in the UK.

## Distinguishing features

- The trumpet-shaped fruiting bodies can vary from blue-black to gray-brown depending on how moist they are.
- The bottom of the cap, that is to say the outside of the fruiting bodies, is blue-gray to gray-black and smooth.
- The bottom part of the stipe is black with a woody consistency and should be cut off.

## Preparing and preserving

The black trumpet suits many dishes either on its own or mixed with other mushrooms, preferably in a stew. The easiest way to preserve it is through drying, but be cautious—dried black trumpet has a very strong flavor, so don't use too much of it when cooking. Soak for 15–20 minutes prior to cooking, and if you remove the water before adding it to your dish, you will get a much milder flavor. The black trumpet when dried and pulverized works extremely well in sauces, soups, and stews.

## Look-alike mushrooms

The black trumpet can really only be confused with the ■ **Ashen chanterelle**, *Crateréllus cinéreus* (see image on the right). Gray in color, it is a very rare species. It has clear ridges on the outside as well as underneath the cap. This mushroom is edible but not common in North America and not known in the UK.

1cm

# Yellow foot

A previously overlooked mushroom, it is now growing in popularity. In areas where it does appear it tends to grow in vast amounts.

Yellow foot is similar to the trumpet chanterelle (see following page) but grows near mossy marshes where other edible mushrooms rarely grow. It is found in rich forests in cooler northern and northeastern regions of the US and adjacent Canada. It seems as if the yellow foot requires more nutrients than the trumpet chanterelle and prefers a soil rich in lime.

## Distinguishing features

- The topside of the cap is red to yellow-brown and slightly scaly.
- The underside of the cap is light yellow to orange and has some *insignificant ridges*.
- The hollow stipe has a similar coloring to the underside of the cap.
- The smell is pleasant and fruity.

NOTE! The yellow foot can at times lack one or more color pigment. This means that one can find smaller groups of entirely yellow or entirely grey-black mushrooms.

## Preparing and preserving

Yellow foot has a nice and mild flavor making it suitable for many dishes. It is most easily preserved by drying. In addition, a glass jar with dried yellow foot in the kitchen is a delightful decoration.

## Look-alike mushrooms

Yellow foot can be confused with the edible mushroom **trumpet chanterelle** (p. 30) but should not be confused with *Leótia lúbrica,* sometimes known as ■ **jelly babies** (see image on the right), that often grow in the same moist and mossy woods but have a gelatinous consistency. The cap is slightly convex with a margin that folds inwards and lacks ridges on the underside. It is inedible but not likely to be poisonous.

# ② Trumpet chanterelle

*Cantharéllus tubæfórmis*

1cm

# Trumpet chanterelle

The trumpet chanterelle has recently become very popular and can be found both in the fresh food aisle in grocery stores as well as in restaurants. It seems to be appearing more and more commonly, which can be explained by the fact that it thrives in woodlands with acidic soil (with low pH). This increases its spread due to the accumulation in acidic soil in northern coniferous forests throughout North America and the UK. The trumpet chanterelle usually grows abundantly within its area. It appears in late fall and is therefore also known as the *winter chanterelle*.

## Distinguishing features

- The topside of the cone-shaped cap can vary from light gray-yellow-brown to brown-black.
- The underside of the cap starts off yellow and later shifts to gray. There are clear, spaced out ridges *in a forked* pattern that partially travel down the stipe.
- The slim stipe is hollow and a brown yellow to yellow, with the base of the stipe being the most yellow.

NOTE! The trumpet chanterelle can lack certain pigments and therefore populations of entirely yellow or gray-black mushrooms can be found.

## Preparing and preserving

Recommended for use in soups and sauces, it also works well fried or sautéed. Not as strong in flavor as the chanterelle but still a spicy mushroom. Ideal for drying. Dried and ground, it works well in soups, sauces, stews, and patés. There have been some cases of people feeling unwell after eating trumpet chanterelle. If you are sensitive, you can blanch the mushrooms first.

## Look-alike mushrooms

The edible mushroom **yellow foot** (p. 28) can be confused with the trumpet chanterelle, but there are no dangerous look-alike mushrooms. However, it has happened that through carelessness or lack of knowledge (see p. 23), people have picked the fatally poisonous ☠ deadly web cap (p. 112-113) while foraging for trumpet chanterelles. They do grow in the same type of terrain but do not resemble each other.

1cm

# Chanterelle

The popular chanterelle has a long season stretching from the middle of June and well into fall. It can be found in a variety of woodland such as beech, mixed woods, pine, and fir as well as in birch woods in the cooler northern regions. (The large picture on the right shows a variety that is commonly known as the "alpine chanterelle").

### Distinguishing features
- The underside of the cap has forked ridges travelling down the stipe.
- It has a raw, sharp taste offering a very unique smell.

### Preparing and preserving
For both the chanterelle and the wood hedgehog (p. 34) frying is recommended. Some suggest that they should be prepared separately, while others believe that the sharp taste works best when the two are mixed together. It is not recommended that you dry the chanterelle, as it can easily become tough and hard during cooking and sometimes a bit too strong in its taste.

### Look-alike mushrooms
One of the reasons that the chanterelle is so popular is probably because it is fairly easy to recognize, even if the picker is uncertain. In the southern US it can mainly be confused with the **"Pale" chanterelle** (see p. 34).

■ **The false chanterelle**, *Hygrophorópsis aurantíaca,* is also very similar to the chanterelle. It's a saprophyte that lives on tree stumps and other decaying woody surfaces and is at times during the fall abundant all over the country. This false chanterelle is known to cause stomach upset in some people, though it is eaten by others without problem. The underside of the cap has thin, narrow gills and the smell is not as strong as the chanterelle. The flesh is thin and soft compared to the chanterelle's firm, almost hard flesh

False chanterelle

1cm

# "Pale" chanterelle

*Cantharéllus pállens*

1cm

# "Pale" chanterelle

This chanterelle does not have a common name in the English language but can be roughly translated as the pale chanterelle and is easy to distinguish from the chanterelle. Despite this, it has taken time for it to gain the status of being its own species, which evidently did not occur until the 1970s. Pale chanterelles grow on tussocks or in small groups in nutrient rich soil and always accompanying leafy trees such as beech, oak, linden, and hazel. It usually peaks slightly earlier than the chanterelle and is sometimes called the *summer chanterelle*. In the US, it has been found in the mountains of North Carolina and vicinity.

## Distinguishing features

that separate pale chanterelle from the chanterelle can be seen on the previous page:

- The pale chanterelle is more robust in size and, when in its infancy, looks more like a plug with a short, thick stipe.
- The top of the cap is pale yellow to *whitish* in color and when handled or in dry weather, yellow patches appear.

## Preparing and preserving

See chanterelles p. 33. Some believe that the pale chanterelle is milder in flavor than the chanterelle, but we believe that they are equal in flavor as well as in their suitability for use in cooking.

## Look-alike mushrooms

The picture below shows a ■ **pallid false chanterelle** *Hygrophorópsis pallid,* which mainly grows in low, dry, grassy areas and is reminiscent of the "pale" chanterelle but softer and thinner. (See a description of the false chanterelle on p. 33). This mushroom is not found in the US.

Pallid
false chanterelle

1cm

1cm

# Wood hedgehog

This mushroom has everything you could ask for from a beginner mushroom. It is tasty, easy to recognize, and lacks any dangerous look-alike mushrooms. It is rarely attacked by maggots and can be found in wooded areas across much of North America and in the UK. In addition, the wood hedgehog has a long season and can often be picked late into fall.

## Distinguishing features

- The whole fruiting body has one color which can vary from pure white via a pale chamois hue, to a pale red-yellowish color. In dry weather the red-yellow color is the most common.
- Underneath the cap, fine teeth can be found attaching to the stipe.

## Preparing and preserving

Wood hedgehog is very similar to the chanterelle and they are closely connected in their classification. Wood hedgehog and chanterelle can be prepared together. It is an excellent mushroom to mix with other mushrooms and very versatile. When dried it has a strong aroma that can leave a slightly bitter aftertaste.

## Look-alike mushrooms

■ **The white hydnum or white hedgehog,** *Hýdnum álbidum,* is a rare species dependent on limey soil. Its outer appearance is very similar to the wood hedgehog. It grows in pine forests (which the wood hedgehog never does) and is not an edible mushroom. This is mainly due to its tough flesh as well as its rarity.

**1 Terracotta hedgehog,** *H. ruféscens* (see image on right), is a closely related edible mushroom which is less meaty. The spikes on the underside of the cap do not attach down the stipe, (compare with the text to the left). It is a good edible mushroom, although older varieties can taste slightly bitter. It can occasionally be found in both deciduous and coniferous forests in the US, Canada, and the UK.

**Albatrellus
confluens
(above)**

1cm

# Sheep polypore

Older mushroom pickers often liken the sheep polypore to an austerity diet as it was used during the war to supplement meat rations. However, for today's youth, many of whom prefer a vegetarian diet, the sheep polypore is an excellent ingredient. The mushroom is a frequent fall inhabitant of the mossy coniferous woods of the northern US and Canada. It is rare in the UK.

## Distinguishing features

- The top of the cap is white but can, depending on age, vary from light brown to gray-white, sometimes with shades of pale yellow-green patches.
- The *pore surface attaches firmly* to the underside of the cap but part ways down the stipe.
- The flesh is firm and white, often with yellow-white hue. Sheep polypore can become green-yellow when heated and, if you are using an iron skillet, can turn a pale black.

## Preparing and preserving

Sheep polypore's firm, meaty consistency makes it extremely versatile. It has an unusual flavor and older mushrooms should be avoided. It is suitable to fry, preferably whole and battered, with onions. Cooked sheep polypore can be ground or mashed in a food processor and mixed into hamburger meat or used in vegetarian dishes.

## Look-alike mushrooms

Sheep polypore can be confused with other fleshy mushrooms that grow on the ground, some of which are inedible. Closely related edible mushrooms are the **Albatrellus confluens**, which grows in similar woodland, and the rarely seen **Albatrellus subrubescens**, that grows alongside pine trees.

2 *A. cónfluens* (see picture above to the left) have a pale yellowy orange color on the cap and often grow in close proximity to each other in North America. The older varieties and those that have been handled can appear with orange patches. Older varieties can have a bitter taste while the younger ones have a mild and pleasant flavor.

2 *A. subrubéscens*, are white with brown patches often appearing in the center of the cap. They are smaller and less fleshy than *A. cónfluens*.

# BOLETES

The Boletes are a large and important family of mushrooms that is simple to learn and recognize. Among this group you can find many good edible mushrooms, very few inedible ones, and a few, but very unusual, poisonous mushrooms, such as one known as the devil's bolete (see following page). Underneath the cap, boletes have horizontal *tubes* that form a pore *surface* that is easy to distinguish from the cap when the fruiting body is mature. The color of the pore surface and the shape and size of the pores vary from species to species and also with the age of the fruiting body.

## Distinguishing the species

When distinguishing between boletes, you should be mindful of...

- whether the topside of the cap (the skin) is slimy, sticky, shiny, dry, matte, or downy.
- the shape and color of the pores.
- the pattern and build of the stipe.
- the flavor of the mushroom when raw. Some of the unusable species have a sharp, peppery, or bitter taste.
- which trees grow in the vicinity. Nearly all boletes live together with a specific species of tree and this is why it is so important to make a note of this.

Unfortunately not only humans appreciate boletes. Deer, squirrels, snails, and last but not least maggots, are big bolete lovers. The larger, more mature fruiting bodies are often riddled with maggots. Always do a basic clean while still in the woods so you don't have to drag several pounds of mushrooms home unnecessarily!

## Preparing and preserving

Boletes are lovely edible mushrooms that can accompany most dishes and be used in all sorts of ways when cooking. They work well sautéed and in stews and can easily be mixed with other mushrooms. The penny bun and other boletes, which are among the most delicious of boletes, are sometimes recommended in older literature as well as foreign guides as raw food. Although they are tasty when raw, there is a risk of raw boletes causing nausea and stomach problems, especially when combined with alcoholic drinks.

Boletes are excellent when prepared in advance (see p. 13) and then preserved, for example by freezing.

Both the consistency and flavor are enhanced through drying, especially when it comes to the velvet bolete, which is much tastier when dried.

## Look-alike mushrooms

■ **Devil's bolete**, *Bolétus sátanas*,
can cause real stomach and intestinal complications but is not deadly poisonous as some older mushroom guides suggest. This mushroom is fairly common in the UK, though in the US the Devil's Bolete has been reported only from California. This poisonous mushroom has a mild and sweet taste.

■ **The bitter bolete**, *Tylopilus felleus* (see description on p. 45), is a look-alike mushroom to both bolete and the brown birch bolete. It is not considered poisonous but if used, has a very bitter taste which will ruin any dish.

Devil's bolete

# ② **Penny bun** *Bolétus édulis*

# Penny bun or king bolete

We will be presenting three types of cep mushrooms. The penny bun, the summer cep and the pine bolete. The penny bun can be found in many regions and in a variety of woodland. It prefers the edge of the forest, roadsides, narrow pathways, and old farms but can sometimes also be found in dense pine forests.

## Distinguishing features

- The skin of the cap is light to dark brown, slightly slimy after rain, and it has an uneven, rugged, or wrinkly edge.
- The flesh of the cap is whitish apart from underneath the skin of the cap where it is more of a *brown-violet* color.
- In the younger mushrooms the pore surface is a gray white color and in the older ones it appears yellow or green-yellow.
- The whole of the stipe, which is usually very swollen, has a light *net veining* or *reticulation,* which can clearly be seen at the top of the stipe, (see image p. 47).
- The taste is mild with a sweet aftertaste.

Some years a parasitic mushroom, **the bolete mold,** can cause extensive damage on parts of the bolete varieties. On those mushrooms afflicted, the cap will stop growing and the stipe will swell to an abnormal size, (see image on right). Eventually the whole mushroom gets covered in a fine coating resembling the mold found on camembert. Obviously these mushrooms are inedible and can cause acute diarrhea. Afflicted mushrooms eventually collapse and become golden yellow and loose, giving off a terrible smell.

*A Penny Bun attacked by bolete mold.*

## Preparing and preserving
See p. 41.

## Look-alike mushrooms
Apart from the closely related edible mushrooms the **summer cep** (p. 44) and the **pine bolete** (p. 46), the penny bun can also be confused with ■ **the bitter bolete,** *Tylopilus felleus,* which has a very bitter taste that destroys any dish if used. (See the description of the bitter bolete on p. 45)

43

# Summer cep

This cep is known as the summer cep as it can be found as early as June and July. It generally grows near beech and oak trees in portions of the eastern US as well as in the UK.

## Distinguishing features
- The cap is smooth and gray-yellow to light brown. In younger mushrooms the cap is covered in a *fine down*. In dry weather the cap can crack into a rectangular pattern. The flesh under the skin is white.
- The pore surface starts off gray-white and then turns yellow.
- The whole of the stipe, which is usually very swollen, has a *light veil* that can be clearly seen from the top of the stipe. (See image on p. 47).

## Preparing and preserving
See p. 41.

## Look-alike mushrooms
The summer bolete is almost identical to *Boletus variipes*, a common edible summer bolete across much of the Eastern US. Apart from the closely related penny bun (p. 42) and the pine bolete (p. 46), the summer cep can also be confused with the bitter bolete.

■ **The bitter bolete**, *Tylópilus félleus*, is the classic look-alike mushroom to the cep family. It has a very bitter taste that ruins any dish if it is added but is not considered poisonous. The bitter bolete has a light brown to chamois colored cap and a pore surface that is first gray turning into a dusky *pink*. On the stipe there is a coarse, dark-veined net. It grows together with pine and other conifers and, at least most years, is very common in forested regions of North America and throughout the UK.

**Pine bolete** *Bolétus pinóphilus*

# Pine bolete

Also known as the *pinewood king bolete*, it is collected in the northern UK. Several similar boletes are collected and eaten in the US.

### Distinguishing features
- The cap is a dark red-brown color (copper-colored).
- Similar to the other ceps the pore surface is at first grey-white and then becomes golden yellow. In this variety the pores can shift in *red-brown* in the older mushrooms.
- The entire stipe, which is usually very swollen, has a light veined net that can be clearly seen at the top of the stipe. (See image on right).
- The taste is mild with a sweet aftertaste

### Preparing and preserving
See p. 41.

### Look-alike mushrooms
Apart from the closely related edible **penny bun** (p. 42) and **summer cep** (p. 44), the pine bolete can also be confused with the

*Pay attention to the notable veined net which is light against a darker background; a typical characteristic for all ceps.*

**bitter bolete**, *Tylopilus felleus*, which has a very bitter aftertaste that ruins any dish when added. (See p. 45 for more information on the bitter bolete.)

47

# Bay bolete

The bay bolete is not especially well known, even if it is just as edible as the penny bun. It is common in the coniferous forests of the northeastern and north-central US and adjacent Canada. It is also seen in forests across the UK. It usually grows with pine trees but can also be found with fir trees and, less commonly, with oak. The bay bolete can appear late in the season.

## Distinguishing features

- The brown cap of the pinhead is lightly downy, then turning bare and, when damp, is slightly sticky.
- The pores are at first light yellow but quickly turn yellow to green-yellow and go a *blue color* when handled.
- The stipe is a mottled yellow-brown color.

## Preparing and preserving

It can be used with most things, similarly to the rest of the cep family. The blue color that appears when you clean and cut the mushroom quickly disappears and the flesh becomes a pale yellow when prepared.

## Look-alike mushrooms

There are no poisonous or inedible look-alike mushrooms to the bay bolete, but it does bear a resemblance to two other closely related edible mushrooms, namely the **1 suede bolete**, *Bolétus subtomentósus,* and the **1 red cracking bolete**, *Bolétus chrysénteron.* The suede bolete has a silky, downy light brown cap and is common in both deciduous and coniferous forests in both eastern and west coast American states. In the older mushrooms the skin of the cap cracks into a reticulated pattern and the flesh nearest the skin of the cap is red. The stipe is also of a red hue. The red cracking bolete lives near leafy trees such as beech and oak.

Bay bolete

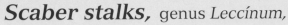

# *Scaber stalks,* genus *Leccínum,*

are easy to recognize and distinguish from other boletes. On the stipes are small scales or small rigid projections that vary in color depending on the variety. Even the top of the cap has a large variety of colors and when considering the cap color you can divide them into two groups:

**1. More or less red in color** (p. 51–53). In this group we consider the **orange birch bolete, the red-capped scaber stalk, orange oak bolete** and **foxy bolete**—all with a pale orange to dark red-brown color on the cap. Collectively these mushrooms are sometimes known as *orange red-capped mushrooms.* The species differ mainly by their location and the color of the rigid projections on the stipes

**2. More or less brown in color** (p. 54–57). In this group we consider the **birch bolete, Leccinum percandidum, Leccinum rotundifoliae, mottled bolete,** and **white birch bolete**. Many of these boletes grow close to birch and are widespread, most commonly in northern areas and especially in the alpine regions.

The flesh is generally light from the beginning but if you snap or cut it, a quick color change occurs. The flesh can go a bluish or blue color or darken to an almost black color. None of these color changes have a negative impact on its edibility.

All brown birch boletes work well when used in food, but it is important to only use firm varieties. This is especially important when using the "brown scaber stalks." In older varieties the flesh of the cap is often loose and the stipe woody.

All scaber stalks have a mild taste and can easily be flavored with herbs or mixed in with more flavorful mushrooms. They are also all suitable to dry.

# ① Orange birch bolete
## *Leccínum versipélle*

The orange birch bolete is commonly found with birch all over the country. In alpine areas it can, together with other boletes, dominate the mushroom population. It arrives early and has its peak season as early as August.

**Distinguishing features**
*Cap color*: brick red to orange colored.
*Pore surface*: grey-beige, darkest in button stage.
*Scabers*: black (even on the button stage).
*Flesh*: when cut is white but quickly goes a blush color and then becomes dark grey.

51

# ① Red-capped scaber stalk
## *Leccínum aurantíacum*

The red-capped scaber stalk grows together with the aspen tree and can be found all over the cooler regions of the US and across the UK. Its season starts early in the whole country, as early as after midsummer in mid-June if there has been enough downpour.

**Distinguishing features**

*Cap color:* orange to brown-red.

*Pore surface:* white, darkens slightly with age.

*Scabers:* pinheads are white, turning red to dark brown with age.

*Flesh:* when cut it is white but quickly blushes to a burgundy before turning black.

# Orange oak bolete
### *Leccínum quercínum*

Reminiscent of the red-capped scaber stalk but grows among oak and beech and is rare in the UK, although not found in North America.

*Cap color:* dark brown-red.
*Pore surface:* white to pale beige.
*Scabers:* dark brown-red.
*Flesh:* white, darkening to black.

# Foxy bolete
### *Leccínum vulpínum*

The foxy bolete grows with pine and is most seen occasionally in cooler regions of the US and adjacent Canada.

*Cap color:* dark red-brown.
*Pore surface:* white
*Scabers:* dark red-brown.
*Flesh:* consistently white.

1cm

# Brown birch bolete

This species grows with birch trees just like many other of the "brown scaber stalks." It is common in the cooler regions of the US and Canada as well as in northern reaches of the UK. In cool boreal forests and the alpine areas it is a common mushroom. Only pick younger specimens, as with the older ones, the caps are often loose in the flesh with a woody stipe.

## Distinguishing features

- The cap can shift through various brown hues but the color is always even and never patchy.
- The pore surface is at first white but later has a grayish tinge, often with *brown patches*.
- The scabers on the stipe are dark brown to almost black.
- The flesh is a consistent white color.

## Preparing and preserving

It has a mild flavor that should be enhanced with herbs and spices.

## Look-alike mushrooms

We would like to warn against confusing it with the ▓ **bitter bolete** (see p. 45) which grows with pine trees and has a bitter flavor. It is sometimes mentioned as a look-alike mushroom to the penny bun, but we believe it bears more of a resemblance to the brown scaber stalks.

Even the **mottled bolete** and the **white birch bolete** (see next page) as well as **leccinum rotundifoliae** are closely related edible mushrooms that can be confused with the brown birch bolete

1️⃣ *Leccínum rotundifóliæ* is uncommon in North America but has been found in Canada and grows together with the dwarf birch. Its cap is light and chamois colored and the stipe is light with white to light brown scabers. The flesh is consistently white.

leccinum
rotundifoliae

1cm

# Mottled bolete

The mottled bolete is one of the best tasting scaber stalks but is still fairly unknown. It grows with birch and along the edges of marshes and is found in Northwest England, especially in the Lake District. It is not found in North America.

### Distinguishing features
- The mottled bolete is easily recognized by its dark cap *with light, irregular patches.*
- There are often blue-green patches at the base of the stipe.

### Preparing and preserving
See p. 50 and brown birch bolete on the previous page.

*Mottled bolete* ready for drying .

### Look-alike mushrooms
In light of its unusual terrain, the mottled bolete only has one look-alike mushroom and that is the white birch bolete, **1** *Leccínum níveum.* It grows in marshy birch woods and the flesh of the cap quickly becomes soft and watery. The young, firm specimens work well when mixed with other mushrooms. The cap can have a pale green hue and, similarly to the mottled bolete, it often gets a blue-green color at the base of the stipe.

White birch bolete

1cm

# ❷ **Slippery jack** *Suíllus lúteus*

# Slippery jack

The slippery jack is a classic edible mushroom. You should, however, only pick the small- and medium-sized fruiting bodies. The cap is usually very slimy in texture and the skin should be removed before being used in cooking. The lower part of the stipe should also be removed as it is usually of a poor quality. The slippery jack is common and widespread across North America and the UK, growing together with white and red pine trees and preferably along dusty pathways and lanes. The season is very long from the summer well into late fall.

## Distinguishing features

- The brown cap is slimy in damp weather and dry and shiny when the weather is dry.
- The pore surface starts off yellow-white and is then covered by a white film, known as a *veil*. When the fruiting body is fully matured, the pore surface is yellow to brown-yellow. The veil then detaches from the margin and forms a ring around the stipe.
- The ring darkens, shrinks, and almost totally disappears in time.

## Preparing and preserving

Once picked, boletes should be taken care of as soon as possible; this is especially important when it comes to the slippery jack. It works very well mixed with other mushrooms and is very versatile.

## Look-alike mushrooms

The slippery jack can be confused with the edible mushrooms **weeping bolete** (p. 60) and **2 larch bolete**, *Suíllus Grevíllei,* as well as the ■ **jellied bolete**, S. flávidus, which, although edible, does not taste good. It has a small slimy ring on the stipe and the pale yellow pores are coarse and angular. It grows with pine trees, in swampy woods, and mires.

*Jellied bolete* to the right.

Larch bolete

1cm

# Weeping bolete

The weeping bolete, or granulated bolete as it is also known, has unfairly had to stand in the shadow of its more "famous" relative the slippery jack. It lacks the fleshy veil that eventually forms a ring around the stipe. The taste is equally good and the flesh usually has a firmer consistency than that of the slippery jack. The weeping bolete grows with pine trees, especially eastern white pines, and prefers soil rich in lime. It is one of the most common and prolific boletes in the northeast US.

## Distinguishing features

- The cap is light brown and in normal weather sticky, in damp weather slimy, and in dry weather shiny and dry.
- The pore surface is at first a whitish yellow and there are often yellow-tinged droplets by the pores. (See fruiting body at the top left).
- The light stipe lacks rings but has, especially at the top, small pail granules that turn brown on the older varieties.

## Preparing and preserving

See the slippery jack on p. 59.

## Look-alike mushrooms

The weeping bolete can easily be confused with the **slippery jack** (see p. 58) as well as with the **2 Jersey cow bolete,** *Suíllus bovínus,* that grow in large quantities alongside pine in Europe and with imported Scott's pine in North America. When viewed from above the weeping bolete and Jersey cow bolete appear the same. If you turn the mushrooms over they are easy to distinguish from each other. The Jersey cow bolete's tubes have a dull, gray and yellow-brown color and the pores are large and rugged. Picking smaller and newly formed Jersey cow boletes is preferable, as the older varieties are not fitting to eat due to their tough texture. It is very suited to drying. (Compare to the velvet bolete on the next page).

Jersey cow bolete

*Dried velvet bolete.*

# Velvet bolete

The velvet bolete is a large mushroom. It grows together with pine in very varied terrain, for example dry, bare rocky areas, mossy deciduous woods, moors rich in lichen, and damp, marshy terrain.

## Distinguishing features

- The velvet bolete has small, *dark scales* and "freckles" on its dry, yellow brown cap.
- The pore surface is somewhat darker in its red brown shade than the cap and stipe.
- When cut or broken the yellow flesh often, although not always, turns *blue*.

## Preparing and preserving

The velvet bolete has previously been discredited as it often turns blue when cleaned but in the past few years has gained popularity. When *dried,* the velvet bolete wins hands down in terms of taste (see photo on the left).

## Look-alike mushrooms

It can be confused with the ② **Jersey cow bolete** (p. 61) and the ■ **peppery bolete,** *Calcíporus piperátus,* which has a strong peppery taste and is therefore unsuitable to eat. The peppery bolete is as a rule smaller and the stipe is thinner with a strong, yellow color towards the base. It also has a delicate consistency. The pore surface, which is usually pale red brown in color, is reminiscent of the Jersey cow bolete in that the pores are large and angular. The peppery bolete lives in both coniferous and deciduous woods and is common across North America and the UK.

Peppery bolete

1cm

# ① **Slimy spike cap** *Gomphídius glutinósus*

1cm

# Slimy spike cap

The slimy spike cap lives with fir trees and is common in boreal forests across the cooler regions of the US and adjacent Canada, as well as in the UK. It is a much loved mushroom, probably due in part to the fact that it is so easy to recognize.

## Distinguishing features

- In damp weather the cap becomes very slimy and in older varieties gets *black patches*.
- The gills are at first white to gray-white and then blacken due to the mature, black spores.
- The stipe is white at the top and *lemon yellow at the base*. In older varieties the stipe darkens.
- In the button stage there is a *crystal clear veil* between the stipe and the cap margin.

## Preparing and preserving

Only pick young fruiting bodies with white or white-gray gills! The chewy film on the cap has to be removed before cooking as does the lower, yellow colored part of the stipe which is slimy and woody. It is recommended that it be mixed with other mushrooms.

## Look-alike mushrooms

There really are no look-alike mushrooms to the slimy spike cap. If viewed only from above, it is possible to confuse it with the slippery jack (p. 58).

*The slimy cap of a **slippery jack** (to the left) and a mature fruiting body of a **slimy spike cap**.*

# Arched wood wax

The arched wood wax prefers to grow in deciduous woods among lichen and moss and preferably in hilly terrain. It is widely distributed in cooler regions of the US and adjacent Canada, growing in association with conifers. In other parts of the country it can be very common. Its season stretches throughout the fall. This species is not known from the UK.

## Distinguishing features
- The dark sooty cap has close-sitting *black streaks* radiating from the middle of the cap to the margin.
- The sparse *waxy* gills are white at first, turning gray with age. The gills travel *downwards* on the stipe.
- Smells of honey.

## Preparing and preserving
It is not recommended as raw food as commonly suggested in older guides. We prefer the arched wood wax mixed with other mushrooms, sautéed or in stews.

## Look-alike mushrooms
The arched wood wax can be confused with similar edible mushrooms such as the olive wax cap and *Hygróphorus olivaceoalbus,* known as the olive wax cap in the US. Both of these edible mushrooms (see p. 69) normally have a slimy cap which the arched wood wax never has.

■ *Club foot, Clitócybe clávipes, has tightly placed gills that attach down the stipe which is swollen at the base. It is common in conifer forests in the northern states and Canada.*

club foot

67

# Herald of winter

The herald of winter appears late fall and grows with pine across the cooler regions of the US and the UK. When the first frost appears the herald of winter begins producing its fruiting bodies.

## Distinguishing features

- In damp weather the *very slimy* cap is golden brown with a gray or olive green hue. In dry weather the cap is dry and shiny.
- The sparse gills and the stipe shift in shades of yellow with orange hues.
- All parts of the fruiting body can have clearly marked *red* patches or streaks.

## Preparing and preserving

Similarly to other slimy mushrooms, the herald of winter should not be picked when it is raining. It works well sautéed and can be mixed with other late fall mushrooms such as the trumpet chanterelle and the conifer tuft.

## Look-alike mushrooms

It can hardly be confused with any inedible mushroom. A few closely related edible mushrooms that also taste good are the  olive wax cap *Hygróphorus olivaceoálbus,* and  *Hygróphorus kórhonenii.* However, they have white gills and the stipes have a dark, *reticular* pattern.

*H. kórhonenii left, and **olive wax** cap have previously been considered to be the same species.*

Olive wax cap

69

# ② Oyster mushroom *Pleurótus ostreátu*

1cm

Indian oyster mushroom

# Oyster mushroom

Cultivated oyster mushrooms have been available in well stocked grocery stores since the 1980s but few mushroom pickers know that it also grows in the wild on stumps and trunks of leafy trees such as maple, beech, elm, poplar, pussy willow, ash, and apple.

The oyster mushroom can plant its fruiting bodies over a long period, as long as there is moisture in the bark. This is a common mushroom across much of the US and southern Canada, appearing as autumn moves into winter and during the winter in southern regions. It is also a frequent mushroom on deciduous trees in the UK.

## Distinguishing features

- The fruiting bodies are found in clusters that consist of a few mushrooms to larger numbers.
- The caps are 3–7 inches (8–18cm) and have a sideways growing stipe. The cap is tongue or kidney shaped, similar to an oyster, and is at first gray-blue, turning gray-brown in the older varieties.
- The gills are sparsely spaced, vary in length, and are white to gray-white in color. They run down the stipe, which is hairy at the base.

## Preparing and preserving

Only pick the fruiting bodies that retain some elasticity in the flesh, and cut away the lower part of the stipe. The oyster mushroom is a good edible mushroom both on its own or together with other mushrooms. It works well dried.

## Look-alike mushrooms

There are other mushrooms in the pleurotus genus that grow on leafy trees. None of these are known to be poisonous and many are good edibles. The oyster mushroom is sometimes confused with the Indian oyster mushroom, ② *Pleurótus pulmonárius*, which is lighter than the oyster mushroom (see photo on left) and is found in the warmer summer months. Otherwise these mushrooms are very reminiscent of each other when it comes to habitat and edibility. The Indian oyster mushroom is common across much of the northeastern US and southern Canada and is frequently seen across the UK.

# ① **Angel wing** *Pleurocybélla pórrigens*

# Angel wing

The angel wing is a mushroom of local interest to mushroom pickers and has been a popular edible in the northwest US and Canada. It grows on dead conifer wood across North America and the cooler regions of the UK.

In the fall of 2004 there were some serious cases of poisoning from angel wing in Japan. The common denominator for those afflicted was old age and several of the victims suffered kidney problems. No other cases of poisoning are known either in North America or in Europe. We therefore don't believe that the angel wing should be avoided. However, it should not be eaten raw and always be cooked well prior to consumption. People who suffer from kidney problems should avoid eating angel wing.

They mainly grow on rotting stumps of conifers and in close proximity to each other but they can also grow on their own.

## Distinguishing features
- The fruiting bodies are at first pure white and then slightly yellow. The mature mushrooms are shaped like ears.
- The stipe is very short or completely absent.
- The white, slightly chewy flesh has a faint but pleasant smell.

## Preparing and preserving
The angel wings short stipe should be removed on picking. We prefer to fry this mushroom. Precooked angel wing can be frozen and also works well when dried.

## Look-alike mushrooms
We deem this mushroom to be so unique that there are no other mushrooms that it can be confused with.

# Wood blewit

The wood blewit is a classic edible mushroom whose value to the culinary world causes mixed opinions. You either enjoy its unique taste or you don't. In some mushroom guides it is recommended that you blanch it first. There are those who don't tolerate the wood blewit very well and thus can get stomach aches if the mushroom is not cooked first. The wood blewits are saprophytes and grow on pine and fir needles, leaves, and grass. It is a frequent inhabitant of compost and brush piles. The fruiting bodies appear in late fall, and it is common across the UK and North America. This mushroom can also be cultivated on composted leaves.

## Distinguishing features

- The cap shifts in shades of brown and gray and very occasionally violet.
- The gills are blue-violet in color, sit close together, and have faint, shallow notches. (See cross section in the photo to the right).
- The blue and violet stipe appears powdery white at the top.
- It has a very unique odor reminiscent of *burnt rubber*.

## Preparing and preserving

The wood blewit can be both fried and boiled. It is less suitable for drying as it has a tendency to become hard and chewy.

## Look-alike mushrooms

■ **The gassy web** cap *Cortinárius tragánus* (see image on the left) and ■ **goatcheese web cap**, *C. camphorátus,* as well as other "purple web caps" can easily be confused with the wood blewit. Commonly these web caps from the genus Cortinarius, as young specimens, have thin, cobweb-like threads called cortinas between the cap margin and the stipe. When the fruiting bodies develop, the threads break and form remnants around the stipe and the cap margin. None of the "purple web caps" are known to be poisonous, but they should be avoided as food. Many grow in deep moss which the wood blewit never does. There are even other purplish web caps that can be confused with the wood blewit. None of these are poisonous but we do not consider them to be useful as food, even if some of them are edible.

1cm

# Pine mushroom

The pine mushroom, also known as the spicy knight, has only recently become popular in the US and Canada. In other parts of the world it has long been appreciated, especially in Japan where it is known as *matsutake* and revered as an edible mushroom.

It grows with pine and hemlock and prefers sandy soil in cooler parts of the US and Canada. The pine mushroom is not found in the UK.

## Distinguishing features
- The cap size is 2–12 inches (5–30 cm) and starts off as a uniform *chestnut* color. Later on it has large brown and flat scales on a lighter base.
- The stipe is often bent, equal in size, and *very rigid*. It is light in color with flattened dark brown bands or patches. In the younger fruiting bodies the gills are covered by a thin *film* that will later form a faint ring around the stipe.
- Its odor is strong and *very unique*, almost sweet.

## Preparing and preserving
It is a firm mushroom with a unique and spicy flavor. One does not do it justice when frying or stewing; rather, it is better off steamed, grilled, or deep fried.

## Look-alike mushrooms
The pine mushroom can be confused with, in culinary terms at least, the uninteresting booted knight, ■ *T. focále* (see below) which is about as common as the pine mushroom and grows in a similar terrain. The color, however, always has a shade of red-brown (sometimes with a yellow-green hue). It also has a thicker ring and a shorter, softer, and (at the base) pointier stipe as well as an odor that is reminiscent of flour. *T. caligatum* is similar but smaller and with a brownish cap. It grows alongside oaks in the US.

booted knight

1cm

# The charbonnier

The charbonnier, also known as the sooty head mushroom, is unknown to many mushroom pickers. It grows late in the season and is hard to discover where it grows among the moss in pine forests. It is, however, an excellent edible mushroom. It is seen late in the season under pines in cooler regions of the northeast US and southern Canada and parts of the UK.

## Distinguishing features
- The cap is unevenly colored in light and dark grays with *black lines* from the center of the cap towards the margin.
- The gills are white at first, but while maturing they shift into a pale *green yellow color*. The same applies to the stipe.
- The white flesh has a smell and taste that is reminiscent of *flour*.

## Preparing and preserving
The charbonnier is a very good mushroom to mix with other mushrooms and works well prepared in a variety of ways.

## Look-alike mushrooms
■ **The ashen knight**, *Tricholóma virgátum,* is the most disagreeable look-alike mushroom,

*ashen knight*    **whitelaced shank**

as it has a very *bitter taste* and is therefore inedible. It lacks the shift of green and yellow on the stipe and gills, and the cap has a *pointy hump* (see image above to the left).

■ **The soapy knight**, *Tricholóma saponáceum,* varies a lot in appearance and some varieties are reminiscent of the charbonnier. The soapy knight does, however, not have the dark lines on the cap and it is also inedible.

■ **Whitelaced shank**, *Megacollybia platyphylla,* has a cap that is reminiscent of the charbonnier with *inward-growing threads.* The stipe is *long and thin, chewy and elastic as well as hollow* and lacks the shift of green and yellow. (See image above to the right). It is an edible mushroom when well cooked, but not highly rated.

1cm

# Parasol mushroom

This towering mushroom grows in pastures and open woodland, and is frequently seen along road banks in forested areas. It is a common mushroom across much of the US and southern Canada and frequently seen across the UK.

## Distinguishing features

- Mature fruiting bodies can be 12–16 inches (30–40cm) in height. The buttons are brown in color. As they grow and mature the outer, brown velum (veil) cracks and forms *darker* scales on a *pale* base.
- The long stipe has a *brown reticular* pattern against a lighter base. A *large ring* can be found on the stipe which can be moved up and down the stipe.

## Preparing and preserving

Caps can be fried whole without batter. The chewy stipe can be cooked with a good stock, or dried and ground to a flour. Ideal for spicing up soups, sauces, stews, and so on.

## Look-alike mushrooms

The parasol mushroom can be confused with the ▓ **shaggy parasol**, *Macrolepióta rhacódes,* which can be found in two varieties: one light garden variety (that is suspected to be mildly poisonous) and a gray wood variety (see image on the right). A common denominator for these varieties is that the stipe is never patterned but completely *uniform* in color.

**shaggy parasol**

In addition the whole fruiting body *blushes* when cut or damaged, and later turns brown. The most dangerous look-alike mushrooms include the ☠ **panther cap**, *Amaníta pantherína*. These mushrooms have *brown* caps with *light patches*, that is to say the complete opposite of the parasol mushroom (see image above). The toxic green-spored lepiota, *Chlorophyllum molybdites*, is the most common cause of mushroom poisoning in central and southeastern US, causing severe vomiting. It is more pale then the parasol, has green spore color, and tends to fruit in grassy areas.

# Shaggy ink cap

There are many varieties of ink caps. Their common denominator is that they grow very fast and have a short lifespan. The shaggy ink cap, or Lawyer's Wig, is the best known and favored as an edible mushroom as well as being very well known for growing around our houses on lawns and in bare soil. It is also frequently found on farms and around livestock barns.

## Distinguishing features

- The cap is at first an elongated egg shape and then becomes bell shaped with white to pale golden brown *scales*.
- The gills are white at first. Starting at the cap margin they then turn pink, then black to finally dissolve in an *ink-like liquid*.

## Preparing and preserving

Only pick varieties with gills that are entirely white and prepare the mushrooms immediately. It works best in stews, and the stipe can be cooked whole and eaten like asparagus.

## Look-alike mushrooms

■ The alcohol ink cap, *Cóprinus atramentárius,* can be classed as a look-alike mushroom to the shaggy ink cap, even if their appearance can vary a lot. The common ink cap has a cap with *lengthways grooves* but is otherwise *smooth* and a silvery gray to brown color.

It has previously been regarded as a good edible mushroom, but new research shows that a substance known as *coprin* can be found in the common ink cap which can have an adverse, long-term effect. It is therefore now seen as a poisonous mushroom. Coprin reacts badly when *combined with alcohol,* something that has given the mushroom its nickname, "the tippler's bane."

common ink cap

# Conifer tuft

The conifer tuft grows on rotting tree stumps, mainly belonging to pine and fir trees. It is a tasty but relatively unknown mushroom for many mushroom pickers. In Sweden it has also been known as *Pale Yellow* hypholoma. It grows in late fall and can be found in large quantities after a timber harvest. Conifer tuft is found across the UK and is widespread in North America.

## Distinguishing features

- The cap is yellow to golden brown with a darker center.
- The gills are at first pale yellow and later on *smoky gray*.
- The stipe is long and thin, pale yellow at the top, and a darker rust color towards the base.
- The flesh is *whitish yellow* and has a *mild taste* which, after a few moments of chewing, is reminiscent of *fresh green peas*.

## Preparing and preserving

Like many mushrooms that grow in tree stumps, this variety is somewhat chewy and has a "woody" stipe that should be removed. The caps can be fried, sautéed, or used in soups; separately or mixed with other mushrooms. The caps also work well when dried.

## Look-alike mushrooms

There are several similar mushrooms that grow on tree stumps and it is important to take a close look at the fruiting body and notice if the mushroom grows on coniferous or deciduous tree stumps.

The sulfur tuft, *Hypholóma fasciculáre* (see image below), is the worst mushroom to be confused with the conifer tuft. It is mildly poisonous and has the following distinguishing features: the gills are at *first sulfur yellow* and *then green-yellow* before finally changing to a darker shade. The stipe is sulfur yellow at the top and the *sulfur yellow* flesh has a *bitter taste*. It grows on dead *leafy* material such as tree stubs and bits of wood and is just as common as the conifer tuft with a slightly smaller spread.

## ② **The gypsy** *Rozítes caperatus*

1cm

# The gypsy

This mushroom grows in conifer and mixed forests across much of the US and Canada but is especially common in cooler forested areas. In the UK, it is known only from the Highlands of Scotland.

## Distinguishing features

- The light brown cap is either powdery white in the center or a *frosty silver gray*.
- The stipe is light golden brown and has a clear ring that *easily detaches*. Above the ring you can see a *lighter* band in an *uneven* zigzag pattern, (see image to the left).

## Preparing and preserving

In older samples, the quality of the stipe is lessened and should be removed around half an inch (1 cm) below the ring. Apart from this the gypsy is very versatile and suitable for drying.

## Look-alike mushrooms

Brown mushrooms with brown gills are not recommended for beginners as they can be mistaken with web caps, but the gypsy is difficult to confuse with any dangerously poisonous variety (see p. 113). There are, however, other web caps that are reminiscent of the gypsy and one of these is the ■ **woolly web cap**, *Cortinárius lániger*. Instead of a large ring on the stipe it has one or two white cotton-like bands that *attach firmly* around the stipe. Like all web caps, the undeveloped fruiting bodies have a *cortina*, or a veil, between the cap and the stipe.

woolly web cap

# Milk caps, *genus Lactárius*

*Agarics,* mushrooms that have *gills (lamellae)* on the underside of the cap, are no doubt the largest group of mushrooms among our woodlands. The agarics that are more easily recognized are *milk caps* and *brittlegills*. The russula varieties have in common a *flesh that is brittle* and crumbles easily when handled (see image below). They are commonly known as "brittlegills."

When fresh, milk caps exude a *colored or clear* fluid when the gills are cut or broken. The cap color varies but usually shifts in brown and grey and the gills are light.

*The photo shows the crumbled fruiting bodies of the "brittle mushroom," that is, varieties of the genus* **milk cap** *and* **brittlegill**, *(see p. 96)*

## Deliciosi

is a group of milk caps that have, as their common distinguishing feature, a milk that is first orange and then turns wine red or green. The *look-alike saffron milk cup* (p. 89) is the most common type and grows in the vicinity of fir trees. The *saffron milk cap* or red pine mushroom (p. 90) are other varieties that can be found and grow alongside pine trees.

### Preparing and preserving
All deliciosi are excellent for eating and can be prepared in a variety of ways. They are not recommended for drying as they can easily become a tad bitter.

### Look-alike mushrooms
The carrot-colored milk makes the deliciosi hard to confuse with any other mushroom. There are other mushrooms that look like deliciosi from the top of the cap, but have white milk and white gills (see p. 91).

# ② False saffron milk cap
*Lactárius detérrimus*

### Distinguishing features
- The cap shifts in the light, and dark orange *zones* become more intense towards the margin.
- Older varieties often have *green patches,* and these do not have an adverse effect on its edibility.

- The *carrot-colored* milk changes after 10–15 minutes to a *wine red* color which, when dry, becomes green.
- The stipe narrows towards the base. It has an *even color* and is porous or hollow.

### Preparing, conserving, and look-alike mushrooms
See page on left for more information.

1cm

# Saffron milk cap

*Deliciosi* is a group among the milk caps, which are described on p. 88. All these varieties have a milky liquid that is first orange and then changes to wine red or green. The *saffron milk cap*, also known as the *red pine mushroom*, lives close to pine trees and prefers a soil rich in lime.

## Distinguishing features

- The uneven shaped cap has concentric circled *orange bands* on a lighter yellow-orangey base as well as the appearance of *grayish belts*. With age, green patches appear on the cap.
- The stipe is often short, hard, and slightly broader towards the base. It has a very characteristic pattern with clearly marked *orange pockets* on a *lighter* base.
- When broken, the flesh is at first carrot colored by the milk but after *an hour* or so darkens to green.

## Preparing and preserving

An excellent edible mushroom. See text on page 88 for more information.

**Lactarius zonarioídes**

## False mushrooms

■ *Lactarius zonarioídes* (see image above), is a very sharp, inedible false mushroom to the saffron milk cap. *The topside of the cap is similar to the deliciosi in that it is zoned,* but once it is picked and broken one discovers a *white* milk and *whitish* gills. It grows with fir trees and is very rare in North America.

# Weeping milk cap

The weeping milk cap is a robust milk cap that is easy to remember due to its unique characteristics. It usually grows with beech, oak, or hazel but can also build mycorrhizal relationships with fir trees. It is a common forest mushroom across the US and southern Canada and in forested areas of the UK.

## Distinguishing features

- This mushroom is large with a cap diameter of up to 6 inches (15 cm).
- The flesh is firm and is rarely attacked by maggots.
- A fresh fruiting body oozes an unusual amount of *white* milk which, as opposed to most milk caps, has a *mild* taste and dries to a tawny brown.
- All parts of the fruiting body become *patchy brown* when handled.
- The mushroom has a strong odor of *shellfish or herring*, similar to the crab brittlegill.

## Preparing and preserving

The weeping milk cap is uncommon in many forests, but its size and strong flavor means one does not need huge amounts to add flavor to a dish. It has a unique flavor which you either love or hate. Its hard texture means it needs to be cooked for at least 20 minutes and it works well in stews and soups. Dill will enhance the shellfish taste.

## Look-alike mushrooms

There are brown milk caps whose coloring are reminiscent of the weeping milk cap but when it comes to size, firmness, and odor it is hard to confuse any of them with the weeping milk cap.

■ **The fenugreek milk cap,** *Lactárius hélvus,* is mildly poisonous and can grow almost as large as the weeping milk cap. However, its milk is *clear* and it has an odor reminiscent of *licorice* or *curry*. It grows in deciduous woods, often in marshy areas and close to pine. In the US, the edible corrugated milk cap (*L. corrugis*) has a darker brown and very wrinkled cap but shares the abundant milk and fishy smell.

Fenugreek milk cap

### "Finnish milk caps"

Rufous, bearded, and northern milk caps are popular in Slavic countries and parts of Scandanavia, but are not commonly eaten in the US or UK. Many consider them to be toxic. They have a sharp taste and must first be *blanched* for five minutes in plenty of boiling water, which is then to be discarded. After this they can be used, but they are mainly recommended for pickling or marinating, making them a nice addition to the more ordinary ways of preparing mushrooms.

## ③ Rufous milk cap *Lactárius rúfus*

The rufous milk cap, sometimes known as the red hot milk cap, is one of the most common milk caps in the boreal forests of the northern US and Canada. It grows with conifers, even in alpine areas such as the alpine birch woods, as well as in bare, rocky areas. It is also found commonly in forested areas of the UK. This mushroom MUST be blanched in boiling water to be edible!

### Distinguishing features

- The cap is dark brown with a low, often pointy hump in the middle.
- The gills are at first almost white, then with aging more of a light ochre (yellow-gray).
- The stipe and the cap are the same color.

### Look-alike mushrooms

The rufous milk cap can be confused with several brownish milk caps but none of them are poisonous except the
- ■ **fenugreek milk cap** (p. 93).

**bearded milk cap**

1cm

## ① Bearded milk cap
*Lactárius torminósus*

Grows alongside birch trees in boreal forests in Northern US and Canada. It is found across the UK.

### Distinguishing features

- The cap margin is, especially with young varieties, clearly *shaggy* and rolled inwards. We believe that the bearded milk cap needs to be blanched twice due to its very sharp taste. Without such blanching, this mushroom is considered toxic in the US.

95

# Brittlegills, *genus Rússula*

Brittlegills are common and many are good edible mushrooms which makes it worth getting to know this family of mushrooms. Brittlegills....

- often have bright colors on the topside of the cap. (Please note that almost no brittlegills are white.)
- have white or yellowish gills.
- are typically brittle and easily crumble (see picture on p.88).
- *never* have a ring or volva around the base of the stipe .

*NOTE! If you wish to know more about brittlegills please refer to one of the comprehensive field guides listed in the back of the book.*

- *never* release a liquid when broken.

As a last option when it comes to determining if the brittlegill is edible or not, you can do the following **taste test**. (NOTE! This is only applicable to **brittlegills**.) Break off a small piece of the gill and chew for a few moments. If...

- the taste (of the raw) brittlegill is *mild or sweet,* or possible *faintly* sharp, it is edible. **Inedible, mildly poisonous brittlegills have a very sharp acrid taste!**

## Preparing and preserving brittlegills

Edible brittlegills can be prepared separately but also work well in all sorts of mixtures with other mushrooms.

The crab brittlegill (p. 104) is the only peppery tasting mushroom among the edible brittlegills. They can usually be preserved in all the usual ways. The copper brittlegill and the darkening brittlegill are less suitable for drying as it gives them a sharp taste. However, in general, drying brittlegills is fine although they can easily crumble.

## Look-alike mushrooms to edible brittlegills

Most of the edible brittlegills can be confused with similar, very sharp-tasting brittlegills containing unknown but mild poisons that can cause light stomach and intestinal problems such as nausea, vomiting, and diarrhea. The brittlegills and milk caps are very popular edibles in Northern and Eastern Europe, though not as highly regarded in the United Kingdom or North America. In many such countries the more acrid species are first blanched in boiling water to render them suitable for food.

# ① Copper brittlegill
## *Rússula decólorans*

Unusually for brittlegills, the copper brittlegill has a relatively even cap color. It is common across northern and mountainous regions of the US and throughout Canada. It grows in coniferous woods and even in alpine areas. During wet summers it can be found as early as July. In the UK, it is seen only in the Scottish Highlands.

### Distinguishing features
- The skin of the cap is sticky when moist and the color is first red-orange then turns yellow in older varieties.
- The flesh is white and turns *crimson* when scraped. In the older fruiting bodies the flesh turns *gray*.
- The taste is usually mild but can, in the younger varieties, be *slightly sharp*.

### Preparing and preserving
Works well in all sorts of dishes. (See drying on page 96).

1cm

# Yellow swamp brittlegill

The yellow swamp brittlegill is widespread across the northern US and Canada and in the alpine areas farther south. It is a common mushroom across much of the UK.

## Distinguishing features

- The skin of the cap is sticky when moist.
- The gills are white at first and then a creamy yellow.
- In the older fruiting bodies the flesh has a gray tinge.

## Preparing and preserving

Mixes well with other mushrooms. It is one of the tastiest edible brittlegills with a sweet, nutty aftertaste. Works well dried.

## Look-alike mushrooms

The yellow swamp brittlegill used to be called the *mild yellow swamp brittlegill* to distinguish it from the *sharper* variety with the same name that is now called the ■ **common yellow brittlegill or the mustard brittlegill**, *Rússula ochroléuca* (see image on right.). The name refers to both the *dirty yellow* color of the cap and its *sharp* taste. The stipe is white but quickly turns a *"watery" gray* with a veined, wrinkly surface.

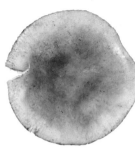

In the beech forests of southern Sweden one can even find the geranium scented russula or bitter russule, ■ *Rússula féllea* (see image on left), which has a dirty, *yellow, straw* color on top of the cap and an acrid sharp flavor. In North America, the almost identical *R. simillima* lacks the geranium scent.

**Common yellow brittlegill**

# Darkening brittlegill

This brittlegill is in many ways reminiscent of the copper brittlegill, (p. 97). It grows in similar terrain and is seen as one of the first brittlegills of the season. In addition, just like the yellow swamp brittlegill (p. 98), its flesh takes on a gray tone in the older varieties. Both the darkening brittlegill and the copper brittlegill have a faint sharp taste that disappears when cooked. These mushrooms are common in cooler regions of the US and Canada but not reported from the UK.

## Distinguishing features

- The cap varies in color from a rusty brown to wine red tones and appears darkest towards the middle. In older varieties it can often appear *faded in patches*. On the buttons, as well as the margins on mature caps, a *grayish film* can be found. The wine red tone is most apparent around the edges. (Compare the image to the right).
- The stipe is white but *gray patches* can appear when bruised or with age.
- The flesh is white at first, turning *gray* and wine red just under the skin of the cap.

## Preparing and preserving

Like the copper brittlegill, this mushroom is most suited when mixed with other mushrooms. Similarly it is less suited for drying (see p. 96).

## Look-alike mushrooms

The darkening brittlegill varies in color and can therefore be confused with other red brittlegills, both edible and mildly poisonous ones. Always taste a tiny morsel of the brittlegill when freshly picked and discard it if the taste is very sharp.

# ① **Baretoothed brittlegill** *Rússula vésca*

1cm

# Baretoothed brittlegill

The cap of the baretoothed brittlegill can vary a lot in its color but is nonetheless easy to recognize. It is also one of the tastiest brittlegills. It grows among grass in the vicinity of birch, beech, and oak trees. It is commonly found in the northeastern US and adjacent Canada and is widespread across the UK.

## Distinguishing features

- Compared to many other brittlegills, the baretoothed variety is unusually *firm and dense*, even when mature.
- The skin of the cap is sticky and earthy and has a unique *salmon pink to pinky brown* hue.
- Furthest out at the margin of the cap there is a minute *white area*, (*edge*), *that the cap does not cover* (hence its name).
- The white gills on the underside of the cap easily become *patchy and brown* in older varieties.
- The stipe is short, firm, and often has rusty brown patches on the somewhat pointy base.

## Preparing and preserving

The baretoothed brittlegill is often firmly attached in the ground. Sometimes you can only see the cap above the surface and as it often grows in loamy soil, it can be difficult to extract. The caps can also get covered in earth, which means they need to be rinsed and rubbed clean. It is recommended that they be mixed with other mushrooms, although with its nutty flavor it is tasty enough to be prepared on its own.

## Look-alike mushrooms

If you learn about the baretoothed brittlegill's many individual characteristics, it is very difficult to confuse it with other varieties.

## Shrimp brittlegill

is a group of brittlegills with the following common characteristics: 1. The light stipe becomes brown when bruised or handled. Somewhere on the stipe there is nearly always a more or less obvious pink to red hue. A stipe with a color combination of *brown and pink* on a lighter base is a typical characteristic for shrimp brittlegills. 2. The smell is reminiscent of *herring or shellfish*. 3. The taste is *mild*.

The shrimp brittlegills are among our favorites in this genus. You can find them in most woody areas and even in alpine areas.

# 3 **Crab brittlegill** *Rússula xerampélina*

It grows with coniferous trees, mainly pine, and is common in coniferous woods across cooler regions of the US and Canada. This is a common species, widespread in the UK.

**Distinguishing features** (apart from those mentioned above) The cap is often *shiny* and a bright to deep red color and darkest, sometimes almost black, in the center. The stipe is hard and a *bright red color*, even right by the gills.

## ③ Entire russula *Rússula íntegra*

# Entire russula

The entire russula, also sometimes known as the nutty brittlegill, grows almost exclusively in *deciduous forests*, often directly through the leaves of older trees. As the cap is somewhat sticky and the fruiting bodies start developing underneath the ground, one can often find traces of earth and needles stuck to the cap. It is an uncommon mushroom in North America and found occasionally in the UK.

## Distinguishing features
- The cap color varies a lot in its hues from deep rust to light brown. Sometimes the colors can be pale, almost *faded and patchy*.
- The gills are fat and high in the mature varieties. Color wise they start off white and then turn pale yellow.
- The stipe is *white*, and occasionally has small *brown patches*.
- The flavor is mild with a drawn-out *sweet aftertaste* reminiscent of almonds or hazelnuts.

## Preparing and preserving
It is recommended these mushrooms be mixed with other varieties. The caps may need to be wiped clean with a sponge under running water. They are suitable for drying.

## Look-alike mushrooms
The entire russula is sometimes confused with *Rússula bádia,* sometimes known as the ■ **burning brittlegill**, which usually grows with *pine trees.* The burning brittlegill has a stipe which sometimes *shifts in red patches* and a *sharp, burning* taste. It is seen in the UK but absent from North America.

burning brittlegill

# Destroying angel

The destroying angel can be deadly poisonous but most people who make it to the hospital in time after ingesting this mushroom will survive without any lasting effects. The destroying angel gives symptoms of poisoning that begin after about six to eight hours.

## Distinguishing features

- The cap is white and, with age, turns yellowish or pale yellow-brown. When damp, the skin of the cap becomes sticky and in dry weather shiny and silky in texture.
- The gills are *always white.*
- The relatively thin stipe may have obvious white scales.
- *At the top* of the stipe is a *thin ring.* Parts of the ring can remain attached on the gills or hang from the cap margin. Often the destroying angel will be attacked by snails, which sometimes consume the ring making it completely absent.
- The base of the stipe is swollen and surrounded by a volva. The base of the stipe is usually well hidden in moss which means that more often than not this distinguishing feature is not included when picking the mushroom.

The destroying angel usually grows in damp and mossy coniferous and mixed tree forests throughout much of the US, Canada, and the UK. It is frequently found in late summer to early fall.

## Warning!

Beginners should completely *avoid picking white mushrooms,* especially those with *white gills.* Confusion between the **common white field mushroom** and the destroying angel can be fatal but easily avoided.

**Destroying angel**
*that is about to leave the pinhead phase. The gills are still covered by a thin film, the partial veil, that later forms the ring at the top of the stipe.*

# ☠ Death cap *Amaníta phalloídes*

# Death cap

When it comes to the level of toxicity and symptoms of poisoning, the death cap is almost identical to the destroying angel (see previous page).

## Distinguishing features

- The cap usually has a greenish hue, sometimes with darker threads growing from the center of the cap towards the edge. The color of the cap can vary, and almost entirely white varieties of the death cap are not unusual.
- A thin skirt-like ring sits high up on the stipe. This is often ragged or partially eaten by snails. The stipe is *mottled*—that is to say, it has a dark pattern against a lighter base.
- The base of the stipe is swollen and surrounded by a thick volva, similar to the destroying angel.

The death cap grows in leafy or mixed woods and sometimes in fields, preferably together with birch, oak, or hazel trees in the UK. Though not native to North America, it has been introduced with nursery stock trees and is now established and common in many regions across the US. Its season is late summer into fall.

## *Warning!*

Very light examples of the death cap have been confused with the **common white field mushrooms**. If comparing only the top of the cap, a death cap is most like a **green brittlegill**, an edible mushroom that is not included in this book. This mushroom, however, lacks a ring around the stipe. In addition, no green brittlegill is swollen at the base of the stipe nor does it have a volva.

*Green brittlegill*     *Death cap*

*An edible mushroom and a deadly poisonous mushroom which are extremely alike when viewed from above. They do however have several features on the stipe which distinguish them from one another.*

111

# Deadly web cap

Web caps make up a large group of mushrooms especially prolific in more temperate forests. Today, no web caps are recommended as edible mushrooms, even if many are relatively safe. Not until the end of the 1970s did *web caps* start to appear in many mushroom guides and one frequent northern species is the deadly web cap. The frightening thing about these mushrooms is that they don't give off any early or significant warnings of someone having been poisoned. Not until three to seven days later (sometimes as long as fourteen days later) do tangible symptoms appear such as increased thirst, tiredness, kidney pain, and at first an increase and then a cessation of the production of urine. In North America, this species has been collected in Canada and is suspected to occur in the northern US. It is occasionally seen in the UK.

## Distinguishing features
- The cap is dull, red brown to yellow brown, and sometimes has elevations but it can also be dome-shaped or slightly flat.
- The gills are also red-brown in color and usually thick, high, and sparse.
- Yellow, irregular rings can be found around the brown stipe.

The deadly web cap grows all over the country but mainly in mossy, fir tree forests as well as pine and beech forests. It seems to have become more common of late.

☠ **Fools web cap**, *C. orellánus*, (see image below), is seen in the UK and craves warm, leafy woods. It is closely related to the deadly web cap and contains the same dangerous poison. On the continent it is more common and has contributed to many serious cases of mushroom poisoning. It is not seen in North America.

Warning!
Beginners should avoid picking brown mushrooms with brown gills. In this group there are many poisonous mushrooms and very few edible mushrooms.

fools web cap

1cm

113

# Mushroom poisoning?

*Marginate Pholiota, also known as deadly skull cap or funeral bell, is a fatally poisonous mushroom with a thin ring around the stipe. It grows on decaying coniferous wood and can be found all over the country, (see below).*

## Symptoms

If you get symptoms such as nausea and vomiting within an hour of eating mushrooms it can be seen as a sign of mild mushroom poisoning caused by **toxins affecting the stomach and intestines,** as their effects occur very quickly.

If after a short time of having consumed the mushrooms, you get other symptoms, e.g. exaggerated sweating, increased saliva or tear production, unusually large or small pupils, or heart palpitations, it can be a sign that you have eaten a mushroom containing a **nerve-affecting toxin.**

The **cell-damaging toxins** are the most dangerous poisons that usually do not give off any symptoms in the first few hours. Poisonous mushrooms belonging to this group are foremost the species described on p. 108–113 as well as the look-alike morel *Gyrómitra esculénta,* also sometimes known as the brain mushroom or elephant ears, (p. 25 and 116) and the skull cap or funeral bell, *Galerína margináta.*

You either feel the symptoms of mushroom poisoning or don't but if you believe you have ingested a poisonous mushroom, contact the nearest **hospital** or the **Poison Control Center** closest to you. If you have remnants of the mushroom or the dish consumed, it is important to save these so that someone with knowledge of mushrooms can try to determine the species. It is important to know which mushroom has been consumed in order to give the correct treatment.

You can even take things into **your own hands**. Initially when suffering from mushroom poisoning try to vomit. The easiest method is to drink a few glasses of water and then force two fingers down your throat, with short nails, as far as possible and move the fingertips until vomiting occurs. If you have activated charcoal (in liquid form) or a similar mixture with active charcoal at home, the dose for an adult is 20–50 ml and for a child 10–25 ml. The charcoal absorbs and neutralizes the poison.

*Broken button of a death cap (below) that shows that a mushroom that is extremely dangerous for humans, can be well liked by maggots.*

## Food Poisoning

Remember that many of those who feel ill after eating mushrooms are not really suffering from mushroom poisoning. They could have eaten a bad meal and simply be suffering from *food poisoning*. Most mushrooms have a short shelf life and should be cleaned and prepared as soon as possible after picking.

> If you suspect that you have been poisoned:
> Call the Poison Control Center at 1-800-222-1222 or 911

# Have mushrooms become more poisonous?

**Brain mushroom,**
Gyromitra esculenta
*(below). A very
poisonous mushroom
that is considered by
people in certain
regions to be a
delicacy.*

Towards the end of the 1970s many known edible mushrooms were reassessed. This was not due to any drastic levels of poisoning but rather that scientists had started to study the chemical makeup of mushrooms. With increased knowledge, some mushrooms and groups of species seemed less appropriate to be classed as edible mushrooms. For example, many clitocybe or blewits, cortinarius (web caps), the brown roll-rim, honey fungus, and morels from the *Gyromitra* family were included. Earlier assessments of the mushrooms nutritional value were based mainly on the authors own experiences and not on any scientific merit. After the big wave of reassessments there has not been a lot of change for the past twenty years when it comes to edible mushrooms.

It is therefore not the mushrooms that have become more poisonous; rather, our knowledge about what they contain has increased! Unfortunately there are misconceptions about toxins in the environment being absorbed by mushrooms that again lead to reassessments of previously accepted edible mushrooms. These assessments have, at least until now, been wrong.

## Individual sensitivity to mushrooms

In the past years we have, during various seminars, exhibitions, and other events, come into contact with people who tell us that they can no longer eat mushrooms they previously enjoyed without any adverse effects. It has often included look-alike morels, chanterelles, and trumpet chanterelles. In these cases it could be due to a simple food allergy, but there can even be other explanations. Forest mushrooms should never be eaten raw; they should be cooked well for a long time. Neither should you eat large amounts of mushrooms at any one time, as foods containing mushrooms can be "heavy" and hard to digest.

A good tip is to exercise caution the first time you eat a new type of mushroom. There are obviously those who are sensitive towards mushrooms, either in general or towards certain varieties. This sensitivity seems to be exacerbated when simultaneously drinking alcohol.

## Heavy metals in mushrooms

*Cd*

Cadmium is a poisonous heavy metal that several types of mushrooms absorb in surprisingly high levels. These mushrooms have a special protein that binds to the cadmium, and it appears that the growth of the mushroom is stimulated by the cadmium. Tests on dry field mushrooms that have been preserved in museums for over a hundred years contain high levels of cadmium. These mushrooms' ability to absorb cadmium is therefore not a new phenomen thought to be caused by recent increases in cadmium, a serious environmental hazard.

# Radioactive cesium in mushrooms

The nuclear power plant accident in Tjernobyl in the Ukraine in the spring of 1986 resulted in several areas of Sweden, in particular the counties of Gävleborg, Västernorrland, and Jämtland, to become contaminated, that is to say affected by a variety of radioactive substances. Eastern winds and a constant rain over several days in these areas was one of the explanations as to why these areas were especially afflicted. Among the radioactive substances it is mainly *cesium-137* that is of some significance. It will remain in the land for several decades to come as the half life is very long.

This means that the land in these areas gives off raised levels of radioactivity. It also means that living organisms in nature that thrive on water and nutrients from the soil will extract and enrich this cesium. This can affect mushrooms and has been shown to do so in various studies conducted both in Sweden and other countries.

Through its mycelium, mushrooms have the ability to absorb a lot of water and minerals from the soil. One of these minerals is potassium, which is reminiscent of cesium in its structure. So when the soil contains cesium, many mushrooms absorb this, especially when levels of potassium in poor soil are already low.

Radioactivity in mushrooms is defined by the measurement *Becquerel (Bq) cesium-137 per unit*, usually per 2.2 lb (a kilo) of fresh produce. The determined maximum level when *selling mushrooms* is 1500 Bq/kg. Recommended yearly consumption in total is 75,000 Bq.

# Methods to reduce radiation in edible mushrooms

If you enjoy picking and eating mushrooms but are concerned about radiation, you can *blanch* the mushrooms before preparing them. The radiation is then reduced by between 60 and 90 percent without affecting the taste. Many of the flavors in mushrooms are not water soluble and are instead released by fats during cooking.

*Directions (for blanching):*

1. Clean the mushrooms and divide into smaller pieces.
2. Place the mushrooms in cold water (1 part mushrooms + 2–3 parts water).
3. Boil the water for around 5 minutes.*
4. Drain the mushrooms of the boiled water (which now contains most of the cesium) and rinse the mushrooms in cold water.

*\* Double blanchi.ng (2 x 2 minutes) reduces the radiation even further.*

There is one other way to reduce the risk of radiation in mushrooms that you intend to cook or preserve.

We have investigated a number of different mushroom caps to see if the various parts of fruiting bodies contain differing amounts of cesium. We have measured the following parts on their own 1) the stipe, 2) the flesh of the cap, and 3) the spore-producing part, that is the spikes, pores, and gills. The result shows clearly that the main amount of cesium in the mushroom can be found in the spore-producing part. Therefore if you remove the spikes, pores, or gills prior to cooking, you can in this way reduce radiation by 40 to 70 percent. This could have an adverse affect on the flavor but we do not have enough information about this currently.

# Comprehensive Mushroom Field Guides

*Mushrooms Demystified, 2nd ed.,* by David Arora, Ten Speed Press, Berkeley, 1986, 959 pages, SUV size, $39.95 paper. ISBN 0-89815-169-4.

*The Audubon Society Field Guide to North American Mushrooms,* by Gary Lincoff, Chanticleer Press, New York, 1981, 926 pages, jacket pocket size, $19.95 vinyl. ISBN 0-394-51992-2.

*Collins Fungi Guide: The Most Complete Field Guide to the Mushrooms and Toadstools of Britain and Ireland,* by Stephan Buczacki.  Harper Collins. $25.00   ISBN-13: 9780007466481.

*Collins Complete Guide to British Mushrooms and Toadstools.* By Paul Sterry & Barry Hughes. Harper Collins. 2009. 384 pages, $26.00  ISBN-13 9780007232246.

# Index

Where there are several page references, the **numbers in bold** indicate the pages where the species and mushroom family are described in more detail as well as pages containing explanations of words and terms.

# How do you avoid mushroom poisoning?

- Never eat any wild mushroom raw.
- Eat a mushroom only if you are 100% certain of its identity and its edible quality.
- When eating a wild mushroom for the first time, eat only a small quantity.
- Only prepare and eat fresh mushrooms. Many mushrooms have a short shelf life and should be cleaned and cooked as soon as possible after picking.

# Edible Mushrooms

presented in this book with words
and images